Project
Succulent

Project Succulent

Genius Ideas for Arranging Succulents, Cacti & Air Plants

BAYLOR CHAPMAN

PHOTOGRAPHS BY PAIGE GREEN

ARTISAN BOOKS | NEW YORK

Library of Congress Cataloging-in-Publication Data

Names: Chapman, Baylor, author.
Title: Project Succulent / Baylor Chapman.
Description: New York, NY : Artisan, a division of Workman Publishing Co., Inc., [2021] | Includes index.
Identifiers: LCCN 2020042944 | ISBN 9781648290329 (hardcover)
Subjects: LCSH: Succulent plants.
Classification: LCC SB438 .C375 2021 | DDC 635.9/525—dc23
LC record available at https://lccn.loc.gov/2020042944

Design by Nina Simoneaux

Artisan books are available at special discounts when purchased in bulk for premiums and sales promotions as well as for fund-raising or educational use. Special editions or book excerpts also can be created to specification.
For details, contact the Special Sales Director at the address below, or send an e-mail to specialmarkets@workman.com.

For speaking engagements, contact speakersbureau@workman.com.

Published by Artisan
A division of Workman Publishing Co., Inc.
225 Varick Street
New York, NY 10014-4381
artisanbooks.com

Artisan is a registered trademark of Workman Publishing Co., Inc.
Published simultaneously in Canada by Thomas Allen & Son, Limited

Printed in China
First printing, March 2021

10 9 8 7 6 5 4 3 2 1

This book has been adapted from *The Plant Recipe Book* (Artisan, 2014)

ALSO BY

BAYLOR CHAPMAN

Decorating with Plants

Home Sweet Houseplant

Tabletop Gardens

To Lila,
for imparting her love
of natural beauty

CONTENTS

THE PROJECTS

INTRODUCTION

Succulents are forgiving and fun to design with (no need to stick with a traditional terra cotta pot) and easy to grow. It's no wonder they have captivated the masses, from newbie plant parents to indoor jungle aficionados. In the following pages you'll see that they thrive almost anywhere, like in old window shutters or tiny Jell-O molds. This versatility, combined with their hands-off care requirements and often friendly, compact sizes, makes them oh so captivating. If you've picked up a copy of this book, they've likely gotten a hold on you already!

To get down to the nitty-gritty, the word "succulent" comes from Latin: *succulentus*, meaning juicy or fleshy. These plants generally adapted from climates with erratic or infrequent water sources. They evolved and honed their skills to store this precious resource into their own private water reservoirs. Some hold water in their leaves, like the thick, waxy aloe, while others have a protective exterior to retain moisture (see the fuzzy panda plants on pages 74–79). Succulents' ability to hold water isn't their only amazing attribute, though: They grow in so many shapes and sizes and fabulous colors and patterns, and some even flower! These characteristics apply to cacti and air plants, too.

In this book, I'm using the term "succulent" broadly and I've included plants that can be considered a semi-succulent or fit the loose criteria of having fleshy, waxy, furry leaves—or any way of holding on to a water source through means other than their roots. Air plants, for example, can utilize moisture in the air. Included here are cacti that are succulents (for not all succulents are cacti), air plants, and bromeliads—a kin to the air plant.

There are several plants that I use over and over and in a variety of combinations. To provide basic care requirements and interesting tidbits, I've spotlighted a few of my favorites. They are listed alphabetically by their genus to help you find the specific plant used in the project. Don't worry too much if you can't find the exact same one—using something of similar size and shape will create a slightly different but just as lovely arrangement. Or bust out and use a plant you think will add color, texture, or an interesting shape. Whether it is an elaborate living centerpiece or tiny single succulent, decorating with plants will make any space feel like home. Dive in, get your hands dirty, and express yourself through beautiful, living art!

The Succulent Arrangement Toolbox

These tools are handy for everyday use with succulents and other plants.

COATED WIRE: To tie soft stems together or to a stake.

DROPPER/SQUEEZE BOTTLE: For watering tiny pots and terrariums.

GLUE DOTS: These help hold cellophane in place.

MISTER: To water air plants gently.

PAINTBRUSHES/SHAVING BRUSHES: Good for dusting plants.

PLASTIC LINER: to protect surfaces from getting wet.

PRUNING SHEARS OR SCISSORS: Used to trim bigger-stemmed plants.

SCREEN: Covers drainage holes to stop roots from plugging up the hole and preventing drainage.

SKEWERS: to create a stem for a succulent cutting.

SMALL ANGLED SNIPS: Angled to prune tiny plants in tiny places.

SMALL SNIPS: Used for pruning delicate plants.

SPOON OR TROWEL: Used to transfer soil or gravel from the bag to the pot.

TURKEY BASTER: Allows you to water in tight spaces.

TWEEZERS: Great for reaching into tiny terrariums to arrange small plants, sticks, and stones.

WATERING CAN: Designed precisely for watering.

WATERPROOF FLORIST TAPE: Used to seal cellophane or foil to make them watertight.

WIRE OR TWINE: Used to tie branches together or to tie a stem to a stake.

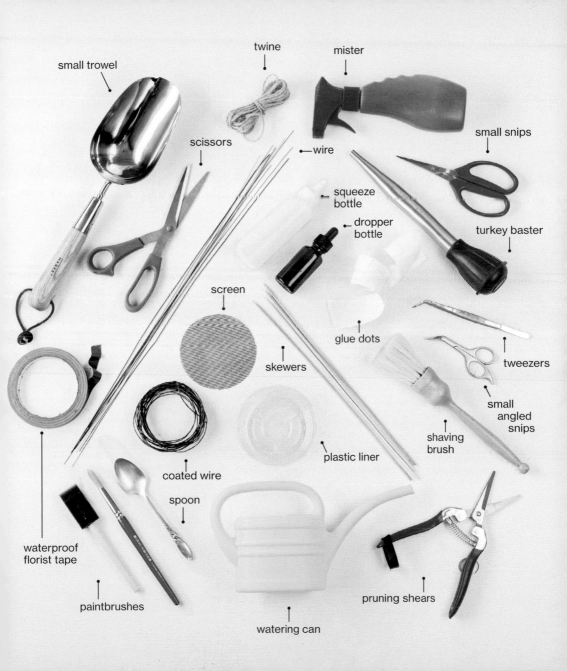

small trowel

twine

mister

scissors

wire

small snips

squeeze bottle

dropper bottle

turkey baster

screen

glue dots

tweezers

skewers

small angled snips

coated wire

plastic liner

shaving brush

spoon

waterproof florist tape

paintbrushes

watering can

pruning shears

Choosing Your Container

Don't be constrained by the notion that plants belong in traditional pots. Look around the house—almost any vessel (and you'll see that term is used lightly here!) can hold a plant. Bowls, cookware, even cups can accommodate container gardens. Pieces of pipe, picture frames, and branches can all be fashioned so that they house gardens, too.

First, consider the growing conditions. The garden will do best and be easiest to care for when the container and the plant(s) in it match up to some extent. Pots with drainage holes are best. This guarantees no swimming pools (and rotting roots) at the bottom of a watertight bowl. Does the plant like conditions that are dry? A widemouthed low bowl will do the trick.

Second, consider size. Will the plant(s) fit in the container? Think too about the needs of the soil and roots along with those of the plant itself. Succulents usually don't mind the squeeze and can be tightly packed in. Air plants can be set side by side and don't even need soil in most cases.

Also, consider scale: how the plant arrangement looks in the pot. To determine if your plant arrangement is properly to scale, you need to consider it in relation to the vessel. As a general guideline, planters should be roughly one-third the height and/or width of your plant. For a more "Zen" design, try the opposite balance.

A patterned basket or pot can be a fun alternative to a traditional vase. Here the U-shape pattern on the fabric basket complements the arrangement by mimicking the shape of the peperomia (Peperomia 'Hope', shown here front and center, with rhipsalis [Rhipsalis capilliformis] to the left and a spear plant [Sansevieria cylindrica] to the right). Just make sure you protect furniture from water leaks with an impermeable liner or dish.

Next, consider color. Sometimes the right color container can make your arrangement sing while another color can make it look drab. For example, the slight hint of yellow in the *Aeonium* 'Sunburst' (page 35) may pop if you place it in a yellow pot, whereas it might disappear in a green one.

Finally, think about the overall look. Is your desired plant design classic or woodsy? A cluster of *Sempervivum* may look more comfortable in a dried date palm than in a flashy neon bowl (although sometimes such contrasts look fabulous—and rules are meant to be broken).

Other times it's best to just let the plant(s) show off. The vertical, spiky legs of the crown of thorns plant, for example, look spectacular when placed in a short clay pot that contrasts the stature and barbs of this unique plant.

WOOD FRAMES are great hosts for plants, whether inside or atop. These are fun to play with, and once plants take root, the frames can even be hung on a wall. They can also be simply set on the table as a cool low, square container.

WOOD BOXES just need to be opened and lined, and away you go with a fun and decidedly unconventional planter.

LOGS AND VASES FROM OTHER NATURAL MATERIALS bring even more of the outside in. Again, these need to be lined to avoid a puddle on the table.

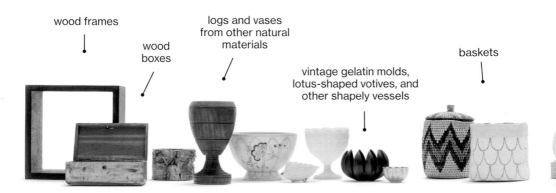

wood frames

wood boxes

logs and vases from other natural materials

vintage gelatin molds, lotus-shaped votives, and other shapely vessels

baskets

VINTAGE GELATIN MOLDS, LOTUS-SHAPED VOTIVES, AND OTHER SHAPELY VESSELS are a chance to have fun and get super creative.

BASKETS feel casual, even a little bit country. Be sure to line them, since they aren't meant to get wet. Some are even flexible and can be stuffed nice and tight with plants.

GLASS TERRARIUMS provide a view from any angle, so be sure the soil and roots look pretty, too. They are best reserved for humidity-loving plants—make sure there's a large opening, and water sparingly.

POTTERY is a classic choice and easy to come by. Most pieces have drainage holes predrilled into them, so be sure to line the inside or set them on top of a plate to protect furniture from moisture.

RUSTIC METAL VESSELS have a charming, weathered, and old-fashioned feel. They can be placed outside, too. Copper and tin, in particular, only look better with time.

PEDESTAL VASES look wonderful with plants that have some drape or droop to them and can add a romantic, even slightly formal, air.

glass terrariums

pottery

rustic metal vessels

pedestal vases

Soil and Amendments

Each container in this book calls for a specific soil type to match the plants and make them happy. In addition to plants and containers, you'll need these things to create your living centerpieces.

SOILS: Soils are mixed with basically the same ingredients but at different proportions, which allows them to hold on to or let go of moisture at different rates. Cactus mix, for example, lets water drain quickly, while potting mix retains water for a bit longer. Some bagged potting mixes contain wetting agents or synthetic materials. Try to steer away from those and stick with organic ingredients. To simplify, the projects in this book call for two types of soil: potting mix and cactus mix.

TOPDRESSING: These pretty additions are a way to top off plant designs and add a bit of polish (they cover the plain old dirt!). The arrangements in this book call for gravel, moss, lichen, and sand, but the possibilities are endless, so get creative! Tumbled glass, buttons, and beads, for example, all work beautifully.

Buying the Perfect Plant

Of course plants are available at garden centers—whose reliably knowledgeable staff and large selection will make the trip worthwhile—but they are also increasingly sold at grocery stores, boutiques, pharmacies, and even pet stores. Online nurseries can be a good option for hard-to-find plants. They, too, can provide a wealth of information. Keep these considerations in mind to ensure that the container garden you create will look its best and last as long as possible:

- Each project in this book provides information to help you locate the right plants, including the scientific and common names. Basic care information, taking into account the specific plant(s) and container, is also provided.

- When in doubt, consult the small plastic tag tucked into the plant's soil to find out how much light the plant wants, how much water it likes, and even if, when, and how often it will bloom.

- Plants can vary greatly in size depending on how and where they are grown. This book calls for a few standard sizes to keep things simple. Some come in round grow pots; some come in square ones. Some will have deep roots and others shallow ones, but if you refer to the size of the grow pot the plant comes in, replicating the effect of each container garden should be a snap. Most of the plants called for in this book are available in 2-inch, 4-inch, or 6-inch grow pots. A few use larger 8-inch and 1-gallon pots. There are a few projects that use succulent cuttings, which are literally the cut-off pieces of a succulent plant.

- No matter the plant's size, always inspect its foliage (leaves). If the plant is supposed to be green, make sure the leaves are actually green, not yellow or brown. Are they upright? Full of life? Avoid plants with wilting or tearing leaves, as well as ones with notched leaves or nibbled bits, both signs of bugs. Check the leaves carefully on both sides for the black or white specks of insects.

- If you're choosing a plant with flowers, look for one with blooms in various stages. Tight buds and blossoming buds promise a future payoff, and full, open flowers give instant gratification.

- Remember, some plants may be tricky to find at certain times of the year. There are succulents, like *Sempervivum*, for example, that are winter hardy (they can freeze), but a lot of succulents are tender and susceptible to cold weather. Though garden centers with greenhouses do allow for a wide selection year-round, not everything is available everywhere all the time.

NOTE: The plants in this book are, for the most part, not edible. Some are even toxic.

Plant Care

Because container gardens are composed of plants that will live together in one place, it's best to choose plants that like the same conditions. While some of the designs in this book break that rule, this is a helpful tip to keep in mind. Even if it is a sustainable composition, you might still get an itch to break up and rearrange your containers, or decide to move plants from container arrangements into a garden.

When combining different plants in a single container, always consider the soil, the water, and the light. Each of the projects in this book lists that information for the main plant and offers tips on how best to care for that specific container garden. Depending on your climate, some plants, like hellebores, can even be transferred to your garden. In their natural habitat, air plants hang from trees in humid, warm climates, like Florida. While succulents might not thrive there—they prefer dry, hot regions—only a few, like *Sempervivum*, can handle freezing temperatures. Because the information provided here is very general, pair this book with a reliable horticulture reference so you can learn how to care for these plants where you live.

NOTE: When arranging or caring for plants, be aware that many plants, not just those with obvious thorns, can irritate skin. People with sensitive skin should wear gloves and everyone should wash their hands thoroughly after gardening, whether indoors or out.

Techniques

Follow these general planting principles for all the container gardens in this book unless the instructions specifically state otherwise.

PREPPING THE CONTAINER

Proper drainage is important for a healthy plant. However, you don't want water all over your table. So be sure to take extra precautions and waterproof the pot, vase, or other container accordingly. This will help keep both the arrangement and its setting—be it tabletop, desktop, or wall—in tip-top shape.

Insert a plastic liner into your container, or line it with special waterproofing aluminum and cellophane that are sold at craft stores. Honestly, though, you can use regular aluminum foil and plastic wrap from the grocery store in a pinch. Both are malleable and will conform to the shape of the container. For extra protection, add a strip of waterproof tape to seal it.

Once your container has been waterproofed, a grow pot or a pot with drainage holes can be safely set inside—remove it to water and then replace. Or you can place plants directly in watertight vessels. Just be mindful when watering to avoid pooling water that will cause rotting roots and sad plants.

Pots with drainage holes need to have those holes covered if the pot will be filled with soil. Simply set a small piece of screen over drainage holes. Not only will this hold the soil in the pot, but it will also assist in drainage by not allowing the roots to block up the holes, so your arrangement can last for months or even years.

As a final waterproofing step, be sure to set pots with drainage holes atop a tray or cork pad. To protect furniture from scratches, place felt dots on the bottom of the tray. Do the same on the bottom of a cork pad, because the plastic layer at the bottom can mark furniture.

Whether planted or just staged (that is, still in their grow pots), plants should sit at the rim of the decorative pot, and sometimes that requires propping them up a bit. I like to use a small upside-down pot or Bubble Wrap. Cardboard, newspaper, and crumpled-up paper towels are not good choices—they will get wet, lose their shape, and start to rot and smell.

PLANTING ESSENTIALS

MEASURE THE DEPTH

Set the plants next to the pot. Are they the right size? Size them up before you plant so you add enough soil. If you will be staging a pot, measuring the depth will tell you how much prop material you'll need.

USE A FUNNEL OR A SCOOP

If you don't have a trowel, a small flexible cup will help you scoop your soil or gravel into the pot. In a pinch, you can also make your own funnel with a piece of paper and some tape.

MASSAGE THE ROOTS

When you unpot a plant, there may be a tightly bound root system. If so, gently massage the roots to give them space to grow. This technique also works if you need to plant in a pot that is slightly smaller than the original.

PLANT AT THE CROWN

Gently make a small hole in the soil and place the plant inside. Fill in with soil around the base, but don't bury the stem. The soil should hit where the stem ends and meets the roots. Roots don't like to be in the air and stems don't want to live under the soil. Gently tamp down the soil.

THE PROJECTS

Aeonium

SOIL Cactus mix

WATER Allow the surface soil
to dry between waterings

LIGHT Bright direct

Some aeoniums are branchy stalks while others grow in a flatter, round form, but the many varieties are prized more for their large succulent rosettes than for their flowers. But when they do bloom, some put on quite a show with a yellow pyramid of blooms. They have a wonderful size range, too, and can hang out of soil for a long time.

Aeonium

On Its Own

PLANT

One 1-gallon *Aeonium*

CONTAINER AND MATERIALS

Glazed pot, 6 inches in diameter and 7 inches tall, with a drainage hole and a flat back drilled for mounting on a wall

One 1-inch square of screen

1 to 3 cups cactus mix

1. Choose a plant size that is close to the size of the decorative pot and complements its color. Set the plant next to the pot; it should be the same height and width. If not, you'll need to make adjustments in step 3. Line the bottom of the pot with the screen and add the cactus mix.

2. Unpot the plant and gently massage the roots to remove any extra soil.

3. Replant the aeonium in the decorative pot, making adjustments to the soil so that the plant sits at the rim of the pot, tipped toward the front. Gently tamp down the soil.

4. Mount the pot on a fence or wall. To water, remove the pot from the wall and water the plant in a sink, allowing the excess to drip through before remounting the pot. Make sure to let the soil dry between waterings.

Aeonium

A Miniature Landscape

PLANTS

One 2-inch dwarf tree aeonium (*Aeonium arboreum* 'Tip Top' is a nice choice)

One 2-inch Mexican snowball (*Echeveria elegans*)

One 2-inch crassula, in bloom (try *Crassula pubescens* ssp. *radicans*)

Two 2-inch bear's paws (*Cotyledon ladismithiensis*)

One 2-inch aloe (*Aloe* 'Christmas Carol' is a good choice)

CONTAINER AND MATERIALS

Recycled glass vase, 4 inches in diameter and 6 inches tall

1 cup decorative gravel

2 cups cactus mix

1. Pour a thin layer of the gravel into the vase (this will help you see any extra water, which you can then pour out by delicately tipping the vase). Add 2 inches of the cactus mix. Unpot the plants from their grow pots and massage the roots to remove extra soil.

2. Plant the aeonium first and then the other plants around its base.

3. Use a spoon to spread the remaining gravel on top of the exposed soil. Let the arrangement dry between waterings, and water until the soil is slightly damp. This arrangement should last between 6 and 12 months.

Aeonium

Woodland Arrangement

PLANTS

One large aeonium with at least three rosettes (*Aeonium* 'Sunburst')

One 6-inch hellebore (*Helleborus*)

One 6-inch rabbit's foot fern (*Humata tyermanii*)

CONTAINER AND MATERIALS

Porcelain bowl, 13 inches in diameter and 8 inches tall

3 cups potting mix

Skewers (if needed)

Coated wire or florist tape (if needed)

1. Select the large aeonium and snip off three rosettes with at least a 3-inch stem. Let the stems dry out (scab over) for a few days.

2. Place the potting mix in the bowl, then unpot the hellebore and set it in the center. If its crown is too low, create a mound with the potting mix.

3. Unpot the rabbit's foot fern and insert it into the open space to the right of the hellebore. Angle it so that it leans over the rim of the bowl.

4. If the stems of the aeoniums are too short, attach a skewer with coated wire or florist tape, or go ahead and skewer the thick stem.

5. Place two of the aeoniums front and center. Place the third, and the largest of the three, resting off the left edge of the bowl.

6. Gently reach into the design and weave the fern fronds through the hellebore. Water lightly. When the hellebore blooms fade and the aeonium cuttings begin to root, disassemble the arrangement and replant it.

Aloe

SOIL	Cactus mix
WATER	Let dry between waterings
LIGHT	Bright direct

The most commonly recognized succulent is *Aloe vera* (shown here), whose gel is used to soothe burns. In these projects, two other, lesser-known varieties are featured. One is more star-like in shape, and the other has fleshy, bumpy leaves. Aloes are easygoing and easy growing, with an intriguing shape to boot.

Aloe

On Its Own

PLANT

One 4-inch lace aloe
(*Aloe aristata*)

CONTAINER AND MATERIALS

Votive vase with open sides

A handful of lace lichen
(*Ramalina menziesii,* shown)
or Spanish moss
(*Tillandsia usneoides*)

1. Choose a vessel that mimics the round and tooth-like shape of the aloe.

2. Soak the lichen to soften it. Unpot the aloe, remove the excess soil, and wrap its roots in the lace lichen.

3. Place the aloe inside the votive, lining up the plant with the "petals" of the vase to show off the repeat pattern.

4. Lightly water the arrangement once a week, making sure there is no standing water.

Aloe

A Desert View

PLANTS

One 4-inch aloe hybrid (*Aloe* 'Pink Blush', *A.* 'Peppermint', and *A.* 'Bright Star' are good choices)

1 lichen-covered stick, about 20 inches long and 1 to 2 inches in diameter

Two 4-inch burro's tails (*Sedum morganianum* or *S. burrito*)

Two 4-inch cathedral windows (*Haworthia cymbiformis*)

Four 4-inch zebra plants (*Haworthia fasciata*)

CONTAINER AND MATERIALS

Copper bowl, 24 inches in diameter

10 to 12 cups cactus mix

2 cups decorative gravel

1. Add the cactus mix to fill the bowl about two-thirds full, and create a slight mound in the center. Unpot and replant the aloe to one side of the bowl.

2. Place the lichen-covered stick in the bowl's center. Unpot and replant the burro's tails so that they drape over the stick. Then unpot and replant the cathedral windows snug against the stick and at a slight angle.

3. Follow suit with the zebra plants. Carefully cover the soil with a layer of the gravel. Let the soil dry out completely before watering about once a week. This arrangement can last a year or more.

Cryptanthus

SOIL   Potting mix amended with peat moss, or orchid or violet mix

WATER   Keep just moist; it wants high humidity

LIGHT   Low to bright

Wavy leaves with distinctive stripes make this terrarium-loving plant look like a starry sea creature—hence its common names starfish plant and earth star. Unlike their tree-loving relatives, this member of the bromeliad family thrives on the forest floor in the wild, so keep them on the moister side. The tiny pups, like babies, are perfect for teeny-tiny arrangements. Just pluck them off and replant.

Cryptanthus

On Its Own

PLANTS

Two 2- to 4-inch earth stars (*Cryptanthus* 'Pink Starlight' is a good choice)

CONTAINERS AND MATERIALS

2 wood block vases

Wax (a candle will do)

1. Look for vases that are about the same size as the plants' grow pots.

2. Melt a candle over the interior of the vases to protect them from moisture.

3. Unpot and plant the earth stars so that the leaves rest on the surface of the vases.

4. Let the plants take root before standing the block vertically; or glue three small wires under the leaves and across the surface of the soil to secure the plants. Keep the plants moist.

Cryptanthus

Starry Kokedamas

PLANTS

One 6-inch bromeliad (*Vriesea gigantean* 'Nova' or *V. Ospinae* var. *gruberi* is a nice choice)

One 4-inch bromeliad (*Neoregelia* 'Donger')

One 4-inch and two 2-inch earth stars (*Cryptanthus* 'Pink Starlight', *C.* 'Elaine', and *C.* 'Ruby' are good choices)

MATERIALS

Branch of Harry Lauder's Walking Stick (*Corylus avenlana* 'Contorta'), about 4 feet long; a curly willow branch will work, too

Two 12-inch squares of sheet moss

6 feet of fishing line

1. Lay the branch in a stable position on a waterproofed surface.

2. Unpot all the plants, removing the pups from the larger earth stars. Create kokedamas by wrapping the soil and the roots of all the plants and pups in moss and securing with string. Place the largest plant (the *Vriesea* 'Nova') first, resting it near the thickest part of the branch.

3. Continue with the smaller bromeliad and the large earth star. Rest the small earth stars and pups on the smallest branches. Mist the arrangement once a day, and remove and soak the plants about twice a week, making sure to add water to the inside of the bromeliad cups. The arrangement will last several months.

Echeveria

SOIL   Cactus mix

WATER   Keep just moist; allow the surface soil
to dry between waterings

LIGHT   Bright direct

This succulent, sometimes known as hens and chicks, offers up juicy rosettes in a variety of colors. If you're lucky enough to find one in bloom, the nodding flowers in sweet bell shapes add height and temporarily change up this reliable, low-growing succulent. Pluck off its offshoots—this one keeps growing, and all you have to do is replant its plentiful pups for more succulents.

Echeveria

On Its Own

PLANTS

Five 4-inch echeverias:
2 painted echeveria
(*Echeveria nodulosa*),
2 shattering echeveria
(*E. difractens*), and
1 Mexican snowball
(*E. elegans*)

Nineteen 2-inch echeverias
or cuttings: 3 *Echeveria*
'Lola', 6 pink-edged or
pink-tipped echeveria
(*E. pulidonis, E. pulidonis*
× *derenbergii*,
E. chihuahuaensis, or
E. 'Captain Hay'), 4 hens
and chicks (*E. secunda*),
4 *E.* 'Dondo', and 2
E. 'Ramillette'

CONTAINER AND
MATERIALS

Painted wood living picture
frame, 12 inches by 7 inches
and 2½ inches tall, with
wire mesh to hold in the soil

4 cups cactus mix

1. Choose plants and a frame color that complement one another.

2. Fill the frame with the cactus mix. Shake the box to even out the soil and allow for it to fall through the mesh.

3. Unpot the echeverias and massage the roots to remove excess soil. Place the plants and cuttings in the frame so that each stem penetrates the soil by at least ⅛ inch. A skewer is helpful with tucking in any roots.

4. Group the larger rosettes together and arrange swaths of like succulents to create a pattern. Let the rosettes rest against the edges so that they spill out of the frame.

5. Enjoy on a tabletop for about a month, until the cuttings and plants have rooted; then you can hang the arrangement on a wall (a regular nail will hold the weight). A felt pad mounted on the wall will help protect it from damage.

6. Let the arrangement dry thoroughly between waterings, and make sure it gets lots of sun.

Echeveria

Elevated Glass House

PLANTS

One 2-inch blue chalkstick
(*Senecio serpens*)

Two 2-inch jeweled crowns
(× *Pachyveria* 'Scheideckeri')

One 2-inch 'Desert Gem'
stonecrop (*Sedum* 'Desert
Gem')

One 2-inch gray echeveria
(*Echeveria secunda* is a
nice choice)

1 lichen-covered twig,
3 inches long

CONTAINER AND
MATERIALS

Teardrop glass, 9 inches in
diameter

1 cup decorative gravel

½ cup cactus mix

5 feet twine

1. Scoop three-quarters of the gravel into the teardrop glass. Tip the teardrop so that the rocks slope up from front to back. Spoon in the cactus mix.

2. Unpot the plants. Begin planting from large to small and back to front, starting with the blue chalkstick, then the jeweled crowns, the stonecrop, and the echeveria. Place in the twig for a nice accent.

3. Use a small funnel to cover the cactus mix with the remaining gravel. Hang the teardrop in a window, using twine. Water with a spoon or a dropper, making sure that none sits at the bottom of the glass.

Echeveria
Fancy and Fragrant

PLANTS

Four 4-inch echeverias (*Echeveria* 'Imbricata' or any other gray one)

One 4-inch earth star (*Cryptanthus bivittatus* 'Pink Starlight')

Two 4-inch coralbells with purple leaves (*Heuchera villosa* 'Palace Purple' or *H.* 'Plum Pudding' is a nice choice)

Two 4-inch ornamental oreganos (*Origanum* 'Kent Beauty')

CONTAINER AND MATERIALS

Vintage silver pedestal vase, at least 12 inches in diameter

2 to 4 cups potting mix

1. Add the potting mix to the vase, creating a mound in the center and leaving several inches from the top of the soil to the rim of the vase.

2. Unpot and replant the echeverias first, mounding them in the center and left of the vase. Tilt the plants forward. Allow the plants at the front to rest on or drape over the rim of the vase.

3. Tilting it forward, place the earth star facing front and center, draping the leaves over the front rim of the vase.

4. Place the coralbells on either side of the echeverias asymmetrically. A view from both sides of the vase will show off their purple leaves.

5. Plant the oregano so that it drapes off one side of the vase and in the center back of the opposite side to create a diagonal line.

6. Mist the earth star three times a week, and keep the oregano moist. When the arrangement begins to droop, disassemble and repot it.

Euphorbia

SOIL	Cactus mix (for the varieties mentioned here)
WATER	Allow the surface soil to dry between waterings; keep dry/light in winter
LIGHT	Bright direct

Euphorbia is one of the largest and most diverse genera around; it even includes poinsettias. Their sap can sometimes irritate skin, so be careful if the milky white goo squeezes out when you cut the stem. Crown of thorns (shown opposite and on page 60) has tiny leaves and sharp thorns but will bloom for months with little care, while a cultivar called 'Sticks on Fire' (shown on page 59) has a smooth, pencil-like form.

Euphorbia

On Its Own

One 4-inch red pencil tree (*Euphorbia tirucalli* 'Sticks on Fire')

CONTAINER AND MATERIALS

Ceramic vase, 3½ inches in diameter and 9 inches tall

1 to 2 cups cactus mix

⅛ cup decorative gravel

1. Pour the cactus mix into the vase.

2. Unpot the red pencil tree and loosen the roots gently to remove extra soil so that it will fit into the vase.

3. Plant so that it is stable and upright with the crown of the plant level with the rim of the vase. Add the gravel.

4. Keep the plant dry and give it all the light you can. The more sun the sticks get, the more red they become. In the shade, they turn green.

Euphorbia

Tabletop Trio

PLANTS

One 2-inch crown of thorns (*Euphorbia milii*)

One 2-inch crassula (*Crassula* 'Springtime')

One 2-inch rat's tail cactus (*Disocactus flagelliformis*)

CONTAINER AND MATERIALS

Ceramic pot, 4 inches in diameter and 3 inches tall

½ cup cactus mix

⅛ cup decorative gravel

1. Fill the container two-thirds full with the cactus mix.

2. Unpot the crown of thorns and carefully place it in the center back of the pot. Use paper to protect your fingers if needed. Then unpot and plant the crassula at the center front.

3. Add the rat's tail cactus in the back to the left of the crown of thorns. Cover the soil with the gravel. Water the arrangement about once a week, tipping the pot gently to make sure no water stands at the bottom. Enjoy for months on end.

Haworthia

SOIL   Cactus mix

WATER   Allow the surface soil to dry
between waterings

LIGHT   Low to bright to bright indirect;
the leaves will become more colorful with more light

Probably the most common variety of haworthia and the easiest to grow is the zebra plant (*Haworthia fasciata*). And for good reason: it has dramatic white lines on the outside of its leaves—although the cathedral window haworthia (*H. cymbiformis*) is almost a clear green. This small plant is slow-growing and easy to keep alive. Like many succulents, it grows its own pups on the side.

Haworthia

On Its Own

PLANT

One 2- to 4-inch zebra plant (*Haworthia fasciata*)

CONTAINER AND MATERIALS

Fluted ceramic vase, 3½ inches in diameter and 3 inches tall

½ cup cactus mix

¼ cup decorative gravel

1. The white stripes are an intriguing attribute of this plant. A white fluted vase is a nice foil to the succulent's horizontal white stripes.

2. Unpot the zebra plant and place it in the vase. Be sure the soil level is just a smidgen below the rim of the vase. If it's too low, add the cactus mix; if it's too high, loosen the roots and try again.

3. Use a paper funnel or a spoon to spread the gravel over the top of the soil. Water sparingly about once a week, and pour out any extra water to avoid soggy roots.

Haworthia

Tiny Terrarium

PLANTS

Three cuttings or 2-inch sedums (try *Sedum dendroideum, S. dasyphyllum,* and *S. burrito*)

Three cuttings or 2-inch houseleeks (*Sempervivum* 'Icicle', *S.* 'Sunset', and *S.* 'Commander Hay' are nice choices)

One 2-inch crassula (*Crassula pubescens* ssp. *radicans*), in bloom

One 4-inch haworthia (try *Haworthia turgida* or *H. cymbiformis*)

One 4-inch zebra plant (*Haworthia fasciata*)

CONTAINER AND MATERIALS

Glass pedestal bias-cut vase, 7 inches in diameter

1 cup decorative gravel

1 cup cactus mix

Two 4-inch-long pieces grape wood

1. Pour three-quarters of the gravel into the vase.

2. Pour in a layer of the cactus mix until it reaches 1 inch below the lower rim of the vase. Place the wood elements so that they form a V shape. Unpot all of the succulents and set them aside.

3. Plant all the succulents as if they were growing out from under the wood elements. Plant in the zebra plant. Use a funnel to get the remaining gravel into all the nooks and crevices. Gently rock the arrangement to even out the gravel. Water evenly with a dropper or a tablespoon until just moist, and allow to dry between waterings.

Hoya

SOIL   Potting mix, perhaps with a bit of perlite
to make it better-draining

WATER   Keep just moist; allow the surface soil to
dry between waterings, especially in winter

LIGHT   Bright

Known also as wax plant and Hindu rope plant, hoya have draping, trailing, fun-shaped leaves. If you're lucky enough to get one in bloom (root-bound plants are said to bloom better), the flowers are amazingly fragrant, too. It's even possible to train it up a trellis.

Hoya

On Its Own

Two 4-inch Hindu rope plants (*Hoya carnosa* 'Crispa Variegata' and *H. c. Compacta*)

CONTAINER AND MATERIALS

Miniature pitcher, 5 inches tall with a 3-inch opening

½ cup potting mix

1. Look for a small vessel with a variety of colors and stripes to mimic the plants' curly and slightly tinted leaves. Fill the pitcher one-half to two-thirds full of potting mix.

2. Thoroughly water then unpot both plants. The roots will likely easily separate. Use only two or three stems of each variety and replant them. Plant the longest stem above the spout for a bit of whimsy.

3. Place the extra stems back in their original containers.

4. Pull the cool, curly leaves down over the rim of the pitcher so everyone can admire their curlicues. Water so it's just moist.

Hoya

Swirls and Spikes

PLANTS

One 4-inch crown of thorns
(*Euphorbia milii*)

One 4-inch Hindu rope
plant (*Hoya carnosa*
'Compacta')

One 4-inch stonecrop
(*Sedum spathulifolium*
'Cape Blanco')

CONTAINER AND MATERIALS

Handmade clay pot
5 inches in diameter and
2 inches tall

One 1-inch square of
screen

Cactus mix (if needed)

1. Add the screen to the pot to cover the drainage hole. Add a shallow layer of cactus mix, if necessary. Unpot the crown of thorns and place it in the back along the tallest wall of the pot.

2. Unpot the Hindu rope plant and place it in front and to the left, opposite the crown of thorns, keeping its leaves upright.

3. Unpot the stonecrop and slip it into the front and to the right, filling the space between the other two plants and draping it over the rim. Water lightly and drain in the sink, letting the arrangement dry thoroughly between waterings.

Kalanchoe

SOIL	Cactus mix
WATER	Allow the surface soil to dry between waterings
LIGHT	Bright direct or low, depending on the plant

The kalanchoe variety known as the panda plant—shown here—does just fine inside. And how can anyone resist those fuzzy leaves? It doesn't just look cool—like other succulents, it also holds moisture so well that you won't need to water it very often. Then there's the dramatic 'Fang' kalanchoe with pointy protrusions on its velvety leaves.

Kalanchoe

On Its Own

PLANTS

One 2-inch panda plant
(*Kalanchoe* 'Chocolate')

Two 4-inch kalanchoes:
1 panda plant (*Kalanchoe
tomentosa*) and 1 feltbush
(*K. beharensis* 'Fang')

CONTAINERS AND
MATERIALS

3 vases made
from copper pipe,
1 to 3 inches in diameter
and 2 to 3½ inches tall

3 cups cactus mix

1 cup decorative gravel

1. Add the cactus mix to the vases until they are about two-thirds full.

2. Match the sizes of the plants to the sizes of the vases, then unpot and plant the kalanchoes.

3. Using a funnel or a spoon, cover the soil with the gravel. Water lightly, letting the arrangement dry thoroughly between waterings and making sure no water stands at the bottom. Enjoy for many months to years.

Kalanchoe

A Planted Pedestal

PLANTS

One 6-inch panda plant
(*Kalanchoe tomentosa*)

One 4-inch eyelash begonia
(*Begonia bowerae* var.
nigramarga is easy to find)

Two 3-inch air plants
(*Tillandsia velutina* is
a good choice)

CONTAINER AND
MATERIALS

Vintage pedestal bowl,
about 12 inches in diameter
and 4 inches tall

One 6-inch square of
sheet moss

1. Select a container that complements the golden hue of the plants.

2. Unpot the panda plant and the begonia. Begin planting with the tallest plant, the panda plant, placing it in the back and to the left of center.

3. Gently place the begonia in the back and right of center so that its leaves drape over the edge of the bowl. Fill in around both plants with the moss.

4. Place both air plants in the front and center.

5. Each week, remove the air plants, soak, shake, and place back in the arrangement. Once a week, lightly water the entire arrangement, letting it dry thoroughly between waterings and making sure water doesn't pool at the bottom. Enjoy for many, many months.

Mammillaria

SOIL Cactus mix or a mixture of perlite,
sand, and soil

WATER Allow the surface soil to
dry between waterings

LIGHT Bright direct

Mammillaria is one of the largest genera in the cactus family. The auburn glow of the spikes make the *Mammillaria bombycina* especially intriguing. But—ouch!—watch out. Some have hooks, so if you do get pricked they are a bit challenging to remove from your finger. When in bloom, this small plant, which reaches only about 2 inches in height, has beautiful little flowers.

Mammillaria

On Its Own

PLANTS

One 4-inch cactus
(*Mammillaria*)

Two 2-inch silken
pincushion cacti
(*Mammillaria bombycina*)

CONTAINERS AND MATERIALS

Serving bowl, 8½ inches in
diameter and 4½ inches tall

Matching ceramic cup,
3 inches in diameter and
3 inches tall

5 cups cactus mix

1 cup white sand

Skewer

Decorative stone

1. Pour the cactus mix to 1 inch below the rim. Unpot and plant the mammillaria and one of the silken pincushions in the bowl and the other silken pincushion in the cup. Use thick paper to protect your fingers if necessary. Add a touch of water.

2. Add a layer of the sand to cover the cactus mix, but don't bury the base of the cacti. Gently shake the vases to even out the sand.

3. Use a skewer to draw a design in the sand. Add the decorative stone. Water about once a week, letting the arrangement dry thoroughly in between.

Mammillaria

Zen Garden

PLANTS

Two 4-inch rabbit ear cacti
(*Opuntia microdasys*)

Three 4-inch silken
pincushion cacti
(*Mammillaria bombycina*)

One 4-inch corncob cactus
(*Euphorbia mammillaris*
'Variegata')

One 4-inch crested cactus
(look for a crested *Opuntia*)

One 4-inch crassula
(*Crassula* 'Morgan's Beauty')

CONTAINER AND MATERIALS

Black walnut box,
12 inches square

Cellophane

4 cups cactus mix

2 cups black sand

1. Line the box with cellophane. To protect the wood completely, make sure the cellophane covers the sides of the box as well.

2. Pour in the cactus mix until it reaches about 1 inch below the rim.

3. Carefully unpot and plant the plants, arranging them close to one another in the middle of the box, as if they were part of a community. Vary the heights and set similar plants at an angle.

4. Pour the sand over the entire top layer. Use a funnel to get close to the plants, but be sure not to bury them. Jostle the box lightly to smooth out the sand. Water very sparingly once a week, and cut the crassula blooms when they fade. Enjoy for many months.

Sedum

SOIL	Cactus mix
WATER	Allow the surface soil to dry between waterings
LIGHT	Bright indirect to bright direct

The burro's tail (*Sedum morganianum*) and its close relative *S. burrito* have fat, succulent leaves that overlap like a thick woven rope and drape over the edge of a container and downward. They are heavy, but be careful: those little leaves detach with the lightest touch. Fret not if some fall off, though, because each tiny fraction will root if placed in soil. Other sedums aren't quite as fragile.

Sedum

On Its Own

PLANTS

One 6-inch and one
4-inch burro's tail
(Sedum morganianum
or *S. burrito*)

**CONTAINER AND
MATERIALS**

Octagonal vase,
9 inches in diameter

2 cups cactus mix

1. Select a planter that tips and turns but stays steady. One with many flat sides allows it to be set at various angles without tipping over. The idea here is to let your sedum spill out as if pouring out of a tipped vase.

2. Spoon in the cactus mix until the vase is one-third full, angling the mix along the shape of the open vase.

3. With extreme care, touching only the soil and the roots, remove the plants from their containers.

4. Place the biggest plant with the longest "tails" in the vase first, at the lowest part of the vase. Plant the smaller, shorter-tailed sedum above it. Gently arrange the tails so that they pour out of the container. Water sparingly, making sure the plants are thoroughly dry between waterings. And resist the urge to move the arrangement around.

Sedum

A Silvery Assortment

PLANTS

Two Sedums: burro's tail (*Sedum morganianum* or *S. burrito*) and Oaxacan stonecrop (*Sedum oaxacanum*)

Nine 4-inch houseleeks (try *Sempervivum tectorum* and *S. calcareum*)

Nine 4-inch echeverias (try *Echeveria* 'Imbricata' and *E.* 'Dondo')

Two 4-inch Catalina live-forevers (*Dudleya virens* ssp. *hassei*)

Two 4-inch pachyphytums (*Pachyphytum hookeri*)

Two 2-inch ghost plants (*Graptopetalum paraguayense*)

CONTAINER AND MATERIALS

Faux concrete bowl, 15½ inches in diameter and 8 inches tall

One 1-inch square of screen

8 to 10 cups cactus mix

1. Cover the drainage hole in the pot with the screen. Add the cactus mix until the pot is about two-thirds full. Unpot and lay out the plants.

2. Start planting in the center of the pot. Choose the biggest plants to go in first. Add more cactus mix, if needed, so that the center plants sit higher than those near the rim of the pot.

3. Fill in with the smaller and draping succulents. Make sure that they are planted level with the rim of the pot. Let the arrangement dry between waterings.

Sedum

Vertical Garden

PLANTS

Five 4-inch stonecrops
(*Sedum rupestre*
'Angelina', *S. acre*
'Elegans', *S. sexangulare*,
S. makinoi 'Variegatum', and
S. reflexum 'Blue Spruce')

Eight 2-inch crassulas
(*Crassula pubescens*)

Four 2-inch houseleeks
(*Sempervivum arachnoideum*
and *S. tectorum* 'Sunset'
are nice choices)

Ten 2-inch echeverias
(*Echeveria secunda*)

CONTAINER AND
MATERIALS

Wood shutter box,
11 inches by 20 inches

One 8-quart bag cactus
mix

2 cups decorative gravel

1. This is a repurposed shutter with a sturdy box built on its back. Drainage holes on the bottom and hooks on the back allow for easy removal and watering.

2. Lay the shutter box flat on its back. Use a trowel and a funnel to pour the cactus mix inside, until the box is almost full. Gently jostle the shutter to allow the mix to settle evenly.

3. Begin with the potted succulents. Unpot them and slip them inside the slats, making sure their soil and roots have fully entered the box. Gently tap down the soil. Next, slip in any cuttings. Make sure they penetrate the cactus mix so that they will root and take hold.

4. Let the shutter lie flat for 1 week to keep the plants in place and encourage them to root.

5. Tip the shutter vertically and, using a spoon, add the gravel to fill in any gaps. Carefully hang the shutter on the wall. Water once a week and let drain, and this arrangement will grow, evolve, and last for years.

Sempervivum

SOIL Cactus mix

WATER Allow the surface soil to dry between waterings

LIGHT Bright direct

Sempervivums are perfect for setting on a windowsill. The rosettes are small enough to rest on the narrow ledge, and they love to soak up the sun. They are sometimes called houseleeks or hens and chicks (yes, just like some species of *Echeveria*). Some varieties grow cobwebs of a weird white film, a trait that's not for everyone but which can be used to great, if eclectic, effect in the right container.

Sempervivum

On Its Own

PLANTS

Seven 2-inch houseleeks
(*Sempervivum* 'Icicle',
S. 'Kalinda', *S.* 'Silver King',
and *S.* 'Rose Queen' are
nice choices)

CONTAINERS AND
MATERIALS

7 vintage gelatin molds
about 2 inches in diameter

1 cup cactus mix

7 tablespoons decorative
gravel

50–50 mixture of glue
(like Elmer's) and water

Decorative platter,
12 inches in diameter

1. Gelatin molds are the perfect size for one perfectly grown sempervivum rosette in a 2-inch pot. Be mindful that some molds have narrow bottoms and therefore might tip over more easily.

2. Unpot each succulent and set it inside a mold. Gently massage the roots if the plant is too tall for the pot—it will flatten out the root-ball to make it fit.

3. Fill in around the edges with the cactus mix if needed, so that the soil fills the container and is nearly flush with the rim.

4. Cover the cactus mix with a layer of the gravel. For extra protection, squirt about a tablespoon of the glue mixture evenly across the top of the gravel in each mold. This will hold the gravel in place should a mold tip over. Set the molds on the platter.

5. Use a tablespoon to water the plants about once a week, and gently tip the molds to release extra water. The arrangement will last for months and months.

Sempervivum

A Date Palm Planter

PLANTS

Nine 2- or 4-inch aeoniums
(*Aeonium* 'Ballerina',
A. 'Sunburst', and *A.* 'Kiwi'
are nice choices)

One 2-inch crassula
(*Crassula pubescens*)

Fourteen 2-inch echeverias
(*Echeveria elegans* and
E. secunda are good options)

Five 2-inch stonecrops:
3 *Sedum dasyphyllum* and
2 *S. rupestre* 'Angelina'

Four 2-inch houseleeks
(*Sempervivum* 'Carmen')

Two 2-inch Cuban oreganos
(*Plectranthus amboinicus*)

Two 2-inch strings of pearls
(*Senecio rowleyanus*)

CONTAINER AND MATERIALS

Dried date palm, with a
split, 56 inches long

5 cups cactus mix

1. Choose a large oval-shaped container with enough room inside for 2-inch plants. If there are no "closers" on the ends, make foil barriers and set them inside. To keep this container steady, you may need to create tiny wood legs. Use a funnel to pour in the cactus mix until the palm is about two-thirds full. Mound the mix in various sections to create rolling hills in the design. Once the plants root, tipping the palm to the side makes it look even more fabulous.

2. Unpot all of the succulents and line them up along the side of the vessel. Play with the design, placing smaller succulents together, allowing single, large succulents to stand alone, adding a blooming succulent, or tucking the smallest succulents in the very narrow edges of the opening.

3. Plant the succulents snugly so that they sit in a tight-knit group. Drape the string of pearls plants over the edge. Let the arrangement dry out between waterings once a week. With proper grooming, this will last for years.

Tillandsia

SOIL   Most need none

WATER   Loves rain, fog, dew, and mist;
soak once a week or mist every few days,
but let dry between waterings

LIGHT   Low to bright

Tillandsia are part of the bromeliad family, but for ease and simplicity they are fittingly called air plants. It's a big category of plants and they come in a range of colors and sizes, from light gray to greens, and from 1 inch tall to 3 feet wide. They often have leaves that arch from a center point, kind of like a star. Spritz with water every few days and soak about once a week. Do not, however, let the plant sit in water for more than a few hours—and make sure it is able to dry within a few hours (for example, don't soak it on a cold, damp evening).

Tillandsia

On Its Own

One 8-inch air plant
(*Tillandsia juncea*)

One 6-inch air plant
(*Tillandsia × floridiana*)

CONTAINER

Porcelain vase, 6 inches in
diameter and 5 inches tall

1. Choose an air plant whose personality will suit the vase, such as those shown here, which resemble the spines of a hedgehog. Match the plant size to the vase size.

2. Place the longer and straighter air plant directly in the vase leaning outward, then set the more curve-shaped one into the vase, angling the plant as needed.

3. About once a week, remove the air plants from the vase, soak, shake, and return.

Tillandsia

A Vibrant Centerpiece

PLANTS

Two 1-gallon echeverias
(*Echeveria* 'Goochie')

One 4-inch pelargonium
(*Pelargonium* 'Oldbury Duet')

Two 4-inch wood spurges
(*Euphorbia amygdaloides*
var. *robbiae*)

Two 4-inch aeoniums
(*Aeonium haworthii* 'Kiwi')

Two 4-inch stonecrops
(*Sedum* 'Cranberry Harvest')

One 6-inch air plant
(*Tillandsia velutina*)

CONTAINER

Vintage pedestal bowl
about 12 inches in diameter
and 4 inches tall

1. Choose a bowl with a pedestal for a formal centerpiece look. Unpot all the plants and remove any extra soil from the roots, breaking up the aeonium rosettes and the stonecrops.

2. Begin by placing the echeverias, one left of center and the other just behind it and in the center, as focal points. Tip the plants toward the front of the bowl.

3. Place the pelargonium on the right edge of the bowl behind the echeverias, and work across the back of the bowl with the wood spurges.

4. Plant the aeonium rosettes and the stonecrops in clusters, leaving a space to the left of the echeveria at the front. Finally, gently place the air plant in that space, resting it on the leaves of the surrounding plants. Drape and fluff all the plants, making sure that the echeverias and stonecrop blooms peek out of the arrangement.

5. This is a temporary arrangement; disassemble after a couple of weeks and replant—the plants will thrive on their own.

Tillandsia

Overgrown Garland

PLANTS

Four 2-inch flaming Katys
(*Kalanchoe blossfeldiana*)

Three 2-inch pink polka dot
plants (*Hypoestes*)

Six 2-inch pink, red, or
fuzzy succulents (*Echeveria*
'Perle von Nurnberg',
E. 'Doris Taylor', or try
a red *Sempervivum*)

Three 2-4 inch red or
blooming air plants (try
Tillandsia ionantha or *T.*
'Houston')

One 4-inch rosary vine
(*Ceropegia linearis woodii*)

Four 6-inch xerographicas
(*Tillandsia xerographica*)

1 clump of Spanish moss
(*Tillandsia usneoides*)

MATERIALS

7-foot length of furniture
webbing

Staple gun

Two 12-inch squares of foil

Wall or large wood piece

1. Spread out the webbing along the wall and staple in the middle. Create loose pockets by stapling the fabric at intervals as you move outward from the center.

2. Cut the foil into small squares. Unpot each plant and wrap the soil and roots in foil. This is a temporary display.

3. Fill the pockets with the plants, tucking one tightly against the other. Slip in a soilless air plant, too. Let the rosary vine dangle down one side and along the center's crest.

4. Gently open the xerographica rosettes to let them clamp onto the fabric. Use a glue dot or wire if extra hold is needed. Separate the clumps of Spanish moss and tuck small bits into nooks.

5. As the plants begin to fade, pluck them out and repot and continue to enjoy the rest of the arrangement. The air plants will likely thrive the longest. Mist one to three times a week; remove and soak the air plants once a week.

ACKNOWLEDGMENTS

Sophie de Lignerolles's talent shines in so many ways—her work as my right hand in many of these creations was, as always, invaluable. Paige Green's talent behind the camera is astonishing. Her sense of light, angle, and composition made the arrangements in this book come to life.

To Kitty Cowles, I can't thank her enough for her introduction, insight, and support. And to Janet Hall, thank you for recommending my design work to Kitty.

Thanks to Lia Ronnen, Bridget Monroe Itkin, and Elise Ramsbottom for their trust, vision, and patience. With their guidance, this book flourished. And thanks to Suet Chong and Nina Simoneaux for their design creativity, and to Carson Lombardi, Keonaona Peterson, and Sibylle Kazeroid for their keen eyes on the copy. Thanks to Molly Watson for her magic with words.

To Lawrence Lee, Robin Stockwell, and SF Foliage for their openness to my plethora of plant questions.

To Thomas Lackey at Stable Cafe, a million thanks for graciously hosting our photo shoots in his lovely café's courtyard and letting us turn his back garden into our photo studio each week. A piece of his garden and generosity is in every photo.

Finally, a shout out to the artisans who created some of the containers for the projects in this book: Heath (page 58), Kelly Lamb (page 88), Esther Pottery (page 73), Pseudo Studios (pages 44, 84), Succulent Gardens (page 50), and Miles Epstein (page 76).

INDEX